CQ-AJU-791

For Ken Mikolowski

A poem is a can of soup
and a lady, too.
Asparagus, curtains, the garbage—
all poems.
Bless this house.

Faye Kicknosway :: Poems & Drawings

ASPARAGUS, ASPARAGUS, AH SWEET ASPARAGUS

The Toothpaste Press
West Branch, Iowa :: November, 1981

Acknowledgements

The drawings owe their parentage to Duchamp, deChirico, Magritte, Ernst, Nanos Valaoritis, pieces & parts of dolls from *The Doll*, an Abrams book, *Mechanical Toys* by Charles Bartholomew, *The World of Toys* by Robert Culff. The poems owe theirs to the second series of 500 poem cards written in 1979 for The Alternative Press.

The publishers thank the National Endowment for the Arts and the Iowa State Arts Council for a Small Press Assistance Grant.

Library of Congress CIP Data
Kicknosway, Faye.
Asparagus, asparagus, ah sweet asparagus.
I. Title.
PS3561.I32A9 811'.54 81-16233
ISBN 0-915124-55-6 AACR2
-56-4 (pbk.)

Contents

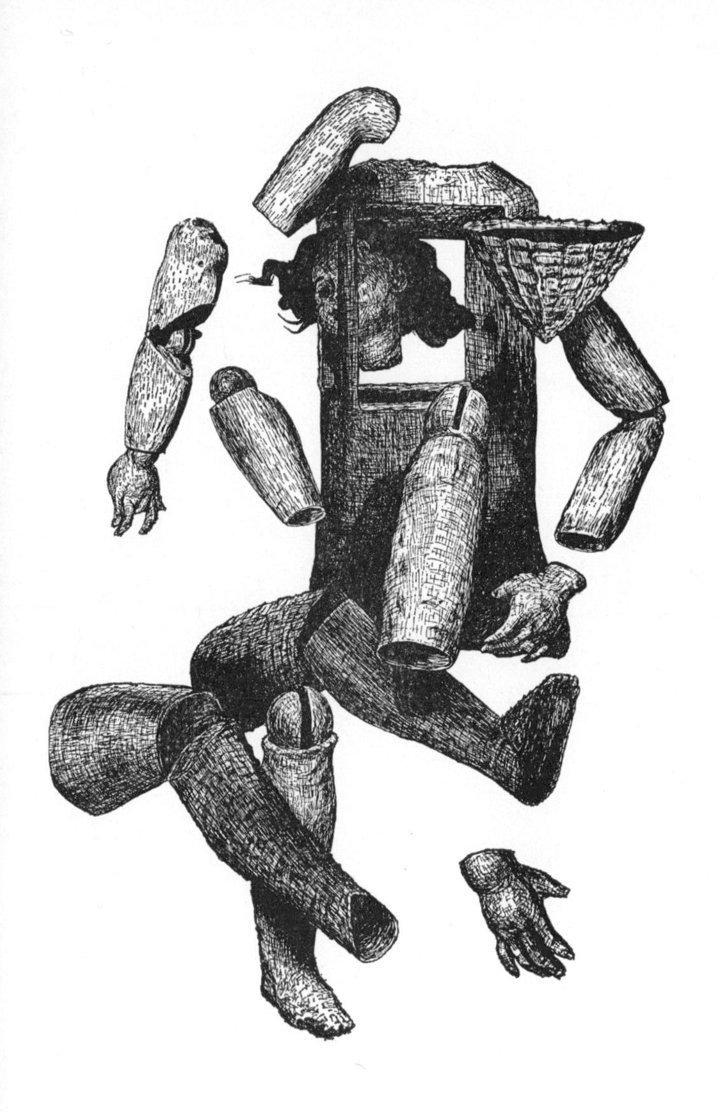

I Invent

1.

I invent a chair:

no legs
no back
no seat
no armrests

It is a very simple chair.

2.

I invent noise
and regret it.

3.

I invent collars, garters,
stays, padded bras,
girdles, tight pants,
indigestion, constipation,
insomnia, and bad breath
all
on the same day.

4.

I invent 'remember' and forget
where it left its gloves
and ask it to go and find them.

I sit down, listen to drawers
open and close. Days pass.
Finally, it brings me its hat,
its tea
and a few comic phrases.

5.

I invent chimneys,
the most difficult chimneys
that begin deep in the earth
where it is too hot
and the bricks keep melting.

6.

I invent the late children
lost between the lawn and the front
door.
I invent their little distracted
faces,
their sunburned, itchy shoulders,
their dance steps and daydreams.
Then I wait for them
to come home.

7.

I invent the boat that is not
on the river
nor on the ground near
the river.

I invent the boat that is inside
the eyes
of the man who sits, legs up,
beside the freeway fence,
watching the traffic.

The Daughter's Song:

It is June because Momma says so.
It is Monday and singing is expected.
I am older because I know it.
I read to the hours so that they stay
quiet
with one another.
Momma calls me, 'Little Monkey',
and Grandma tells me
that my footprints bleed
into the sand
in her egg timer.

List #1

Potatoes
Peanut Butter
Cauliflower
Rice
Tofu
Plums

Haircut
Dentist
Wash Car
Bank

List #2

The grocery sack
The waist tie
The stocking
The thing suspended
The daughter with the soap
The neighbor
with the keys

List #3

To have known the wagon
To have paused beside it
To have been muscular
Fatigued
To have lain down
Properly attired
To have been seen
To have been constantly aware

Tofu

If you are Chinese,
and are natural in your habits
and graceful in your speech,
you might be thrown off
a boat,
or dangled upside down like bait
in a pond
so that a fat, short emperor—
pen in hand—
may write a simple,
but profound
poem
about the sound
water makes
touching the shoreline.

SHORT STORY:
The Rings and Donkeys and Sand

Once there was a small flea
who loved to sing Italian opera
while sunbathing in between the hairs
of a small German sheep dog
asleep on the Côte D'Azur.
There was another flea, his brother,
who lived a miserly, niggardly life
on the back of a Mexican Hairless.
Their sister Roxanne, unsatisfied
with her ordinary life, ran away
with a flea circus and,
after leading a long, horribly
dissipated life, died
an anonymous death
in the tail
of a black Pekingese.

The Naked Man

1.

The naked man wears a hat
because he is worried about his hair.
It escapes by wagon,
by dog sled and camel
out the window.

2.

He sucks his moustache, glares
and is not happy
being naked,
but he is afraid bees
will fly up his shirt sleeves
and sting him, or up his pant legs
and sting him. Clothes
are perilous; one takes one's life
into one's hands
wearing them.

3.

The naked man is annoyed
and stands in the corner,
calmly, ceremoniously.
He will not get dressed because
there is no other light
in the room save the one
shining through his skin.

4.

This is divided into 'wet' and 'dry'.
He has taken his underwear off,
his moustache.
He looks at his toes
through his foliage:
"Look at those toes," he says.
"I'm going to wash them."

5.

Sing a song to the naked man:

-x-x
-x-x-
-x-x/-x-x-x-xx
-x--xx-x--xx
xx-xx--xx---x

6.

The naked man looks in the bathroom.
There is a spider in the bathtub.
He yells for the children
to catch it in toilet paper
and weeps bitterly
when they refuse.

7.

He looks in the mirror.
Why is his hair so wet?

It is too close to the light bulb.
He throws his clothes out
the window.
It comforts him.

8.

I invent a nightgown
for him.
He has his pants off
and is attracted.
I hand it to him.
He hangs it on his fingers,
on his moustache.

9.

He ties his fingers into pretty bows
behind his back
and walks pigeon-toed
through the house,
simping through his moustache:
"I'm the little old lady
who cares for you;
I'm your mother."

10.

She was rude.
He was naked, singing.

Poem

They sit
until they're fat, and then
they die
taking it all
with them.

Natural History, Part 1.

1.

The man at the center of this lettering
has an unusual face. His smile,
his unblinking eyes, don't look at them,
look instead
at his hands: one is clawed
and the other is human. It is not hair
covering him, nor is it fur,
it is the letters of the alphabet
which have been beaten
with short, spatulat sticks
into a fine dust; he has rolled
in it.

2.

Water enters,
scowling and impatient.
It has no spiritual justification
for its bad manners.
When it sweats, the air dies.

3.

(Hear the faucet?
It remembers.
And the onion in the sink, it
remembers, too.)

4.

In the beginning, the earth was hollow.
A can of soup rose up
from its center. Pick axes, prayer,
the weather, nothing
could destroy it.
It grew larger than the earth,
finally splitting the earth
in half. The halves revolved
in twin orbits. The can burned
and fattened. Clouds came.
Mountains sprang up.

Natural History, Part II.

What has cardboard to do with celery?
Coffee with living on the third floor?
Which season am I?
Was I ever twelve years old
in Roanoke, Virginia?
Did my mother rag curl my hair
and dream of orange juice and 1943?
Was I the virgin who thought sex
was eating too much spaghetti
and throwing up?
Which sock

do I put on first in the morning?
And vegetables?
And wallpaper?

The Poetical Notes of Hyacinth Beemer

1.

The day is hard.
It is marked by bathers
and bath water. By the shuddering
of flesh. How solemn the expression
on each face. A few herbs rubbed into
the skin and the heart beneath
trembles.

2.

She was a little woman with eyes like reeds,
and she blinked them open,
blinked them shut.
Her mouth was a round, black coin
above her chin.

3.

She seems to float between the walls.
Small, her edges are chipped and broken.

There are holes in her, windows.
She is like a root vegetable
and leaks moisture onto the table.
Light comes from her, and she grasps
things

in her hands, becoming a skin
around them.

4.

She could be called 'Mole', although
she despises the earth,
and digs in it
and rolls on it only
to divide it. She weeds the bottoms
of plants
and brushes all corruption
from their tiny roots, giving them
room to complicate
themselves.

From The Pumpkin To The Shoe

First, it was a juicy chair; she took a bite.
The lamp, coarse and old, complaining,
complaining, leaned from its cord
toward her.

The entrails of birds penetrated the night;
each thin shadow dissolving into a noise
like oars
being lifted from water.

The window was quiet, open, deliberate
in its breathing. Its fingers were at angles
on the sill
and its long, bushy tail hung down
the outside wall.

She was not satisfied and braided her fingers
together, assuming a posture she would abandon
month by slow month.

She was like a tunnel : tremulous, alive.

There was a meticulous love
in the air, the color of garnet,
and it moved
across the walls as she despaired
of the roses
and peonies that grew
in wild profusion

along her memory's old,
wooden fence.

The day's attention dragged
and ended. She wept bitterly
and slept
in a tidy little knot, her skirt
a sea
that crowded and disobeyed her

In Mysterious Ways

Condoms keep catching at the river's
skirt. And some girl, swimming
in New Jersey—a virgin who doesn't know what
it's all about, has no explanation
for anything—gets bitten
on her trunks by one of them
that's filled out
like a little fish. And it's a Miracle
a *real* Miracle,
what pops her belly out.
But no one
believes her; no one.

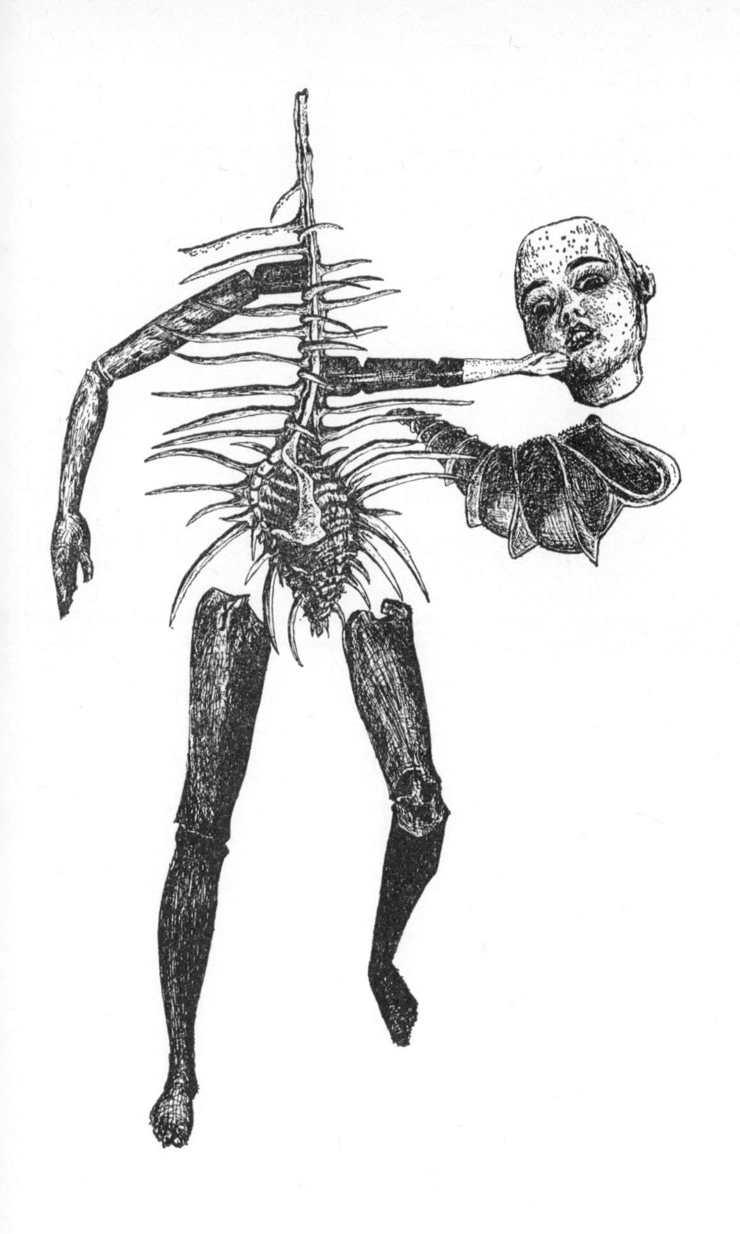

Primer

A.

Which of these relevant metaphors
support? :

1. The men, unexplained, were blooming.
2. She was a field and he a tent.
3. Excellence, the weeping it made
   on the floor.
4. Because the air ran untidy to the tree.

B.

Directions for Exercise 1 :

1. My heart is a thick watch.
2. At noon, the underwear.
3. The ready coat and stocking
   left singing in the tub.

C.

The Sonnet, the Ode, the Rondel,
all are acquired. Rhyme them
with variety.

D.

Watch and Describe:

a. —the coat torn open, the letter gone.
b. —there is a steady wave, a bright

cry
from an upstairs window.
c. —nine small holes in the tree.
d. —frail, her deeds are hidden,
most kindly, in her heart.

E.

The missing words are tonics poured into
the lap.
The missing words suds up when you forget them.

F.

I sits me down
I wears me out
I turns me sorely
Like I was meat on a spit
Above a fire
I sends me home
My momma scolds me
And I feels helpless, soured
By it all

G.

Choices (a) and (c), possibly
(d) and (b), but not
in that order.
Metrically attractive and singular
in appearance are (e)
and (g).

H.

The newspaper is sinister.
She digs in the garden, smokes
the flowers out.

I.

-x/-x/--/-x/-x
--x--x-x---x-
--x--x--xxx-
--x/-x/--x/-x
x-/x-/--

("Is it about a hyacinth?"
"Skip it.")

J.

Metrically:

1. the young kingdom

2. Sally had a dress
3. booming, in the rushes
4. the frolic of beautiful Mary

K.

Shall we, coldly and idyllically,
rush to name ourselves
as 5-line stanzas or terza rimas?
What original foot shall we use
as antecedent?
And the ballad form, its possible density,
and great, structural beauty? *

* The passage concludes with a few
remarks of a thoughtless nature.

Synthesis

THESIS:

The hills wear
a rind of sleep
above the valley.

The river drifts,
mercilessly. Indolent and sighing,
the wind loosens
its robes:
the long, graceful sleeves.
How sad

the moon's face, white
and pierced
by branches.

ANTITHESIS:

Today, the air is like mold
on something left
in the refrigerator
too long.

The window, mouth, nose
dare not open
toward it.
How it watches, so still,
ready to jump.

West Grand Boulevard

Three small children,
impatient at the bus stop,
practice karate:
a toe to the kidney,
to the throat.
Their mother, her arms full
of packages, watches them.
She's going to give them
something more old-fashioned
when they get home.

Poem

A roof. A sky made of water.
Trees that belch like frogs.
Shall I remember later
What it was like today?

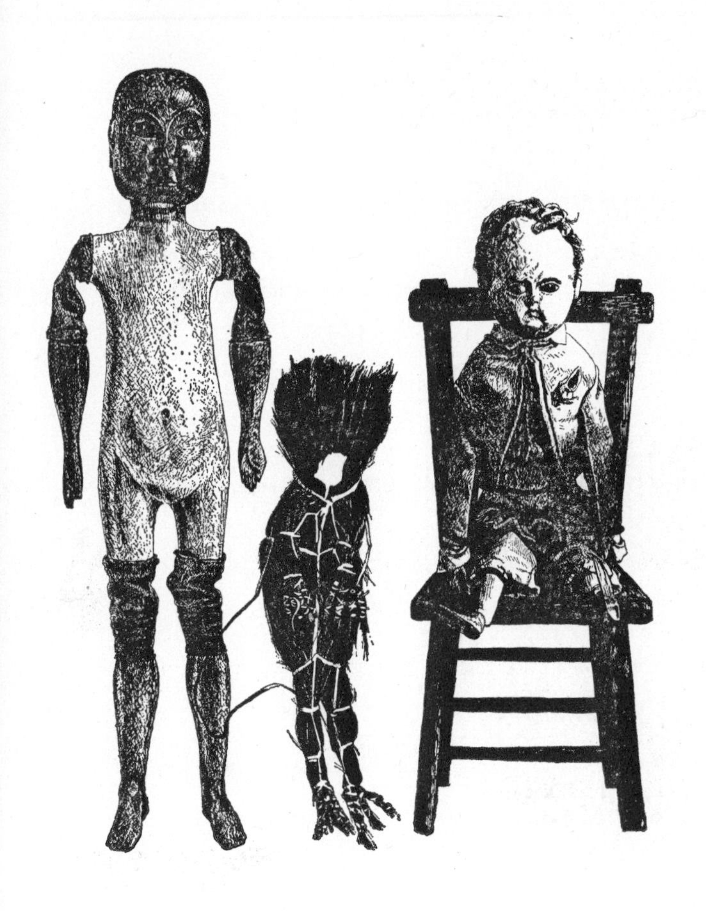

After Hilary, age 5

1.
Fat men
eat horrible food
like toads & gravel
&
sit in the garage
like laundry
until your Aunt Doris
makes them leave.

2.
The woman who lives in Grandma's
china cabinet
eats large cats
and little children
between 1 and 4 p.m.
every Thursday.
Grandma lets her
because of the alligators
in the radiators
and the turtles
in her throat.

3.
If you spit at the Moon
it will lay eggs in your hair
and your mother will be gooseflesh
and your father will die

The Advice

Vacancy; stop near it.
What falls from it
is important.
The essentials are always
the same: hardship, deprivation.
Be dazzled by what looks dark.
Always avoid warmth
and ask that which opens
to close.

Book Report

1.

King Lear was a lively person.
He had a talent for vile
and swollen prose and never spoke
in a simple, human voice.
He had daughters, most delicious
in their clothes, who spoke constant
poetry.

2.

His one daughter, the good one,
had nice, clean hair.
That's how he always knew her:
by her nice, clean hair. The other daughters
could have been anyone, he'd never
have guessed. The only way he knew
them was they never called him 'King'.

3.

He sent each daughter the following
letter:

"Will you tidy up the dust in this room?
Will you obey it? sit down near it
and wink your eyes at it?

"I am your driver's license, I know all

the facts
and possibilities of your Name.
The possession of anything larger
than your Name
would be a joke."

4.

He quit his job as king
and took a larger job
as a crazy person. He liked the job
a lot
and had a friend
who gave him lessons.

5.

If King Lear had been more simple,
if he had done his own laundry
and taken out his own garbage, he'd never
have had the problems he had.
Most of what happened to him
was made up, and he didn't like it.

6.

He took his friend into the woods
and yelled at him, and at the storm
that blew his clothes off.

No wonder his daughters took offense.
They wanted him to act normal,

not clumsy and peculiar.
He wanted to be memorable and walked
around the countryside
yelling at the trees and houses.
No wonder it all blew down or
fell over. No wonder his friend

got sick, or hanged, or shot.
On the way to the emergency room,
his friend whispered to King Lear
that he should stay alive
and start a war.

7.

When his friend died,
King Lear no longer had anyone
to whisper jokes up his sleeve
and his daughters found him
sitting in a tree, helpless.

He died that night.
He did it long-winded and surly,
but that was alright;
it was done.

The Unattended Foot

But again, the children would not,
nor the chimney, the bacon,
the pebbles; they all went barking

toward the bright green wagons
that played in the high
young trees.

There was a handkerchief,
a foot unattended.
There was a life, a forearm, a nickel
left. The smile looked down
from the calendar,
but not its teeth.

Could it fly up?
Could it fall down, or button
or explain?
Could it weep? It sang, and in American:

"O chestnut eyes, I wish you leaves
and twigs. Do not wake here, o wisely
quiet and baffled stranger; the tenderness

that pries
and slowly climbs
is not attractive. Sleep well
in clean boots. Divine the door
and wait

until dark and passionate hands
possess the house.

What hair,
what biscuits blow?
How stretch yourself,

how rest in mystery
and turn it
clearly seen?"

Colophon

Italian Olde Style type handset by Ellen Weis. Designed and printed by Allan Kornblum. 1,150 copies on Curtis Tweedweave, smyth sewn and glued into wrappers by Rebecca Henderson at Prairie Fox. 100 copies on Ragston, numbered and signed by the author, cased in cloth by Constance Sayre at Black Oak Bindery.